UNDER THE MICROSCOPE

SCIENCE TOOLS

9 SCIENTIFIC CLASSIFICATION

John O.E. Clark

GROLIER
EDUCATIONAL

About this set

SCIENCE TOOLS deals with the instruments and methods that scientists use to measure and record their observations. Theoretical scientists apply their minds to explaining a whole range of natural phenomena. Often the only way of testing these theories is through practical scientific experiment and measurement—which are achieved using a wide selection of scientific tools. To explain the principles and practice of scientific measurement, the nine volumes in this set are organized as follows:

Volume 1—Length and Distance; Volume 2—Measuring Time; Volume 3—Force and Pressure; Volume 4—Electrical Measurement; Volume 5—Using Visible Light; Volume 6—Using Invisible Light; Volume 7—Using Sound; Volume 8—Scientific Analysis; Volume 9—Scientific Classification.

The topics within each volume are presented as self-contained sections, so that your knowledge of the subject increases in logical stages. Each section is illustrated with color photographs, and there are diagrams to explain the workings of the science tools being described. Many sections also contain short biographies of the scientists who discovered the principles that the tools employ.

Pages at the end of each book include a glossary that gives the meanings of scientific terms used, a list of other sources of reference (books and websites), and an index to all the volumes in the set. There are cross-references within volumes and from volume to volume at the bottom of the pages to link topics for a fuller understanding.

Published 2003 by Grolier Educational, Danbury, CT 06816

This edition published exclusively for the school and library market

Planned and produced by Andromeda Oxford Limited,
11-13 The Vineyard,
Abingdon, Oxon OX14 3PX

Copyright © Andromeda Oxford Limited

Project Director Graham Bateman
Editors John Woodruff, Shaun Barrington
Editorial assistant Marian Dreier
Picture manager Claire Turner
Production Clive Sparling

Design and origination by Gecko

Printed in Hong Kong

Library of Congress Cataloging-in-Publication Data

Clark, John Owens Edward.
  Under the microscope : science tools / John O.E. Clark.
      p. cm.
Summary: Describes the fundamental units and measuring devices that scientists use to bring systematic order to the world around them.
Contents: v. 1. Length and distance -- v. 2. Measuring time -- v. 3. Force and pressure -- v. 4. Electrical measurement -- v. 5. Using visible light -- v. 6. Using invisible light -- v. 7. Using sound -- v. 8. Scientific analysis -- v. 9. Scientific classification.
  ISBN 0-7172-5628-6 (set : alk. paper) -- ISBN 0-7172-5629-4 (v. 1 : alk. paper) -- ISBN 0-7172-5630-8 (v. 2 : alk. paper) -- ISBN 0-7172-5631-6 (v. 3 : alk. paper) -- ISBN 0-7172-5632-4 (v. 4 : alk. paper) -- ISBN 0-7172-5633-2 (v. 5 : alk. paper) -- ISBN 0-7172-5634-0 (v. 6 : alk. paper) -- ISBN 0-7172-5635-9 (v. 7 : alk. paper) -- ISBN 0-7172-5636-7 (v. 8 : alk. paper) -- ISBN 0-7172-5637-5 (v. 9 : alk. paper)
  1. Weights and measures--Juvenile literature. 2. Measuring instruments--Juvenile literature. 3. Scientific apparatus and instruments--Juvenile literature. [1. Weights and measures. 2. Measuring instruments. 3. Scientific apparatus and instruments.] I. Title: Science tools. II. Title.
  QC90.6 .C57 2002
  530.8--dc21
                    2002002598

About this volume

Volume 9 of *Science Tools* looks at various examples of scientific classification. First to be described is the Periodic Table, the chart that classifies more than 100 chemical elements. Rocks and the minerals they are made of also have classification schemes. Among living things there are various ways of classifying plants and animals, as there are for bacteria and human blood groups. Finally, we look at the wide variety of heavenly bodies: planets, moons, comets, stars, and galaxies. They also have their own systems of classification. Classification is itself a science tool: It helps us understand how things interact and how they relate to each other.

Contents

Main units of measurement

Scientists spend much of their time looking at things and making measurements. These observations allow them to develop theories, from which they can sometimes formulate laws. For example, by observing objects as they fell to the ground, the English scientist Isaac Newton developed the law of gravity.

To make measurements, scientists use various kinds of apparatus, which we are calling "science tools." They also need a system of units in which to measure things. Sometimes the units are the same as those we use every day. For instance, they measure time using hours, minutes, and seconds—the same units we use to time a race or bake a cake. More often, though, scientists use special units rather than everyday ones. That is so that all scientists throughout the world can employ exactly the same units. (When they don't, the results can be very costly. Confusion over units once made NASA scientists lose all contact with a space probe to Mars.) A meter is the same

length everywhere. But everyday units sometimes vary from country to country. A gallon in the United States, for example, is not the same as the gallon people use in Great Britain (a U.S. gallon is about one-fifth smaller than a UK gallon).

On these two pages, which for convenience are repeated in each volume of *Science Tools*, are set out the main scientific units and some of their everyday equivalents. The first and in some ways most important group are the SI units (SI stands for Système International, or International System). There are seven base units, plus two for measuring angles (Table 1). Then there are 18 other derived SI units that have special names. Table 2 lists the 11 commonest ones, all named after famous scientists. The 18 derived units are defined in terms of the 9 base units. For example, the unit of force (the newton) can be defined in terms of mass and acceleration (which itself is measured in units of distance and time).

▼ Table 1. Base units of the SI system

QUANTITY	NAME	SYMBOL
length	meter	m
mass	kilogram	kg
time	second	s
electric current	ampere	A
temperature	kelvin	K
luminous intensity	candela	cd
amount of substance	mole	mol
plane angle	radian	rad
solid angle	steradian	sr

▼ Table 2. Derived SI units with special names

QUANTITY	NAME	SYMBOL
energy	joule	J
force	newton	N
frequency	hertz	Hz
pressure	pascal	Pa
power	watt	W
electric charge	coulomb	C
potential difference	volt	V
resistance	ohm	Ω
capacitance	farad	F
conductance	siemens	S
inductance	henry	H

▼ **Table 3. Metric prefixes** for multiples and submultiples

Prefix	Symbol	Multiple
deka-	da	ten (×10)
hecto-	h	hundred (×10^2)
kilo-	k	thousand (×10^3)
mega-	M	million (×10^6)
giga-	G	billion (×10^9)

Prefix	Symbol	Submultiple
deci-	d	tenth (×10^{-1})
centi-	c	hundredth (×10^{-2})
milli-	m	thousandth (×10^{-3})
micro-	µ	millionth (×10^{-6})
nano-	n	billionth (×10^{-9})

Scientists often want to measure a quantity that is much smaller or much bigger than the appropriate unit. A meter is not much good for expressing the thickness of a human hair or the distance to the Moon. So there are a number of prefixes that can be tacked onto the beginning of the unit's name. The prefix milli-, for example, stands for one-thousandth. Therefore a millimeter is one-thousandth of a meter. Kilo- stands for one thousand times, so a kilometer is 1,000 meters. The commonest prefixes are listed in Table 3.

Table 4 shows you how to convert from everyday units (known as customary units) into metric units, for example from inches to centimeters or miles to kilometers. Sometimes you

may want to convert the other way, from metric to customary. To do this, divide by the factor in Table 4 (not multiply). So, to convert from inches to centimeters, *multiply* by 2.54. To convert from centimeters to inches, *divide* by 2.54. More detailed listings of different types of units and their conversions are given on pages 6–7 of each volume. You do not have to remember all the names: They are described or defined as you need to know them throughout *Science Tools*.

▶ **Table 4. Conversion** to metric units

To convert from	To	Multiply by
inches (in.)	centimeters (cm)	2.54
feet (ft)	centimeters (cm)	30.5
feet (ft)	meters (m)	0.305
yards (yd)	meters (m)	0.914
miles (mi)	kilometers (km)	1.61
square inches (sq in.)	square centimeters (sq cm)	6.45
square feet (sq ft)	square meters (sq m)	0.0930
square yards (sq yd)	square meters (sq m)	0.836
acres (A)	hectares (ha)	0.405
square miles (sq mi)	hectares (ha)	259
square miles (sq mi)	square kilometers (sq km)	2.59
cubic inches (cu in.)	cubic centimeters (cc)	16.4
cubic feet (cu ft)	cubic meters (cu m)	0.0283
cubic yards (cu yd)	cubic meters (cu m)	0.765
gills (gi)	cubic centimeters (cc)	118
pints (pt)	liters (l)	0.473
quarts (qt)	liters (l)	0.946
gallons (gal)	liters (l)	3.79
drams (dr)	grams (g)	1.77
ounces (oz)	grams (g)	28.3
pounds (lb)	kilograms (kg)	0.454
hundredweights (cwt)	kilograms (kg)	45.4
tons (short)	tonnes (t)	0.907

The need for classification

There are more scientists today than ever before, gathering more data and information than have ever existed. To help impose order on this sometimes bewildering accumulation, each branch of science has its own methods of classification.

Some of the most important science tools are ones that cannot be physically handled or plugged into the electricity supply. These tools are ideas: ways of thinking about things logically—"scientifically." The proper organization of information is one example. To see how this works, we will do an experiment. We need no apparatus, just our brains. This is what scientists call a thought experiment—another useful kind of science tool.

▼ **This classification scheme** divides 15 common small creatures into categories according to their physical appearance. Any other similar minibeast could be found a place in the scheme merely by counting its legs and wings.

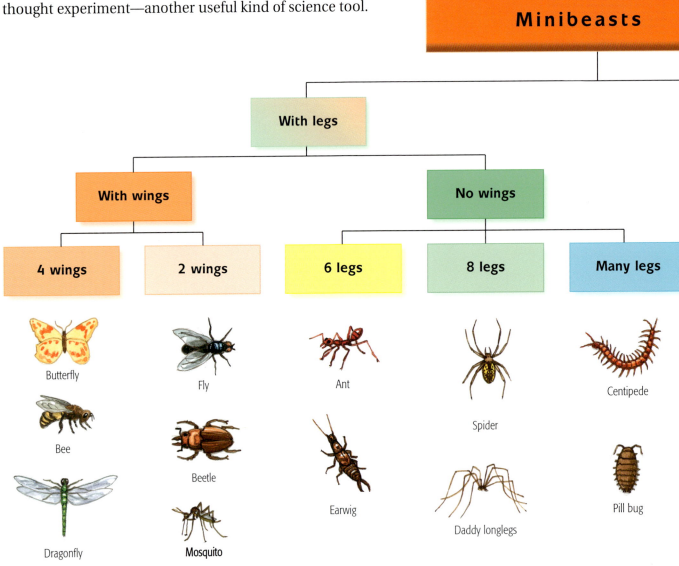

Box of Minibeasts

With legs

With wings — **No wings**

4 wings | **2 wings** | **6 legs** | **8 legs** | **Many legs**

Butterfly

Fly

Ant

Spider

Centipede

Bee

Beetle

Earwig

Daddy longlegs

Pill bug

Dragonfly

Mosquito

▲ **If you come across this beetle** (actually a very rare hermit beetle), you might not see its two wings because, as for all beetles, they have hard outer coverings over them that are moved aside when the beetle takes off.

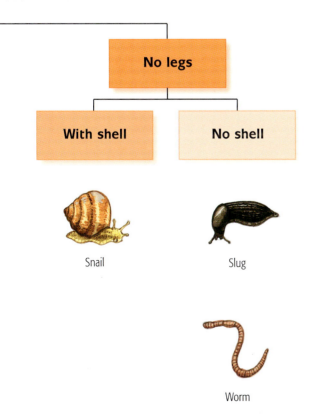

Bug hunting

To begin this thought experiment, imagine you are going into the backyard, a park, or a wood. The experiment has two aims. First, you are going to collect as many small creatures as you can—bugs, spiders, worms, and so on (but no small mammals, such as mice). One advantage of only thinking about it is that you do not have to handle the creatures! The second aim is to come up with a scientific way of organizing your collection.

Just think of as many minibeasts as you can, and imagine you have collected one of each in a box. Now to organize the collection. There are many ways of going about it, but first of all split the collection in two, with minibeasts that have legs in one group and legless minibeasts in the other. There are probably not many in the second group, perhaps a slug, a snail, and a worm. We can subdivide this group into ones with a shell (the snail) and ones without a shell (the slug and the worm), as on the right-hand side of the illustration here.

What about the other, larger group, the ones with legs? A further division would be to put those with wings in one subgroup and those without wings in another, as on the left-hand side of the illustration. Winged creatures can be divided yet again into those with four wings and those with only two wings. Minibeasts with no wings go into one of three groups: those with six legs, those with eight legs, and those with many legs. Again, you can see the results in the illustration.

Classification schemes

Just by thinking about it, we have come up with a way of classifying an imaginary collection of small animals. It is not how a zoologist would do it, but it does correspond in places to the correct biological classification. For example, all the six-legs are insects, and all the eight-legs are arachnids (spiders and their relatives). But all the winged creatures are also insects. And an earwig does actually have wings; but because you can't normally see them, the earwig has turned up in the wingless category. A beetle's wings are also normally hidden, but we put the beetle in with the two-wings because we know that beetles can fly—we've seen ladybugs (which are beetles) flying from plant to plant.

Still, our classification scheme is scientific, and we can fit into it other creatures we did not "pick up" first time around. How about a moth, for instance? It goes straight into the four-wings group. So would a wasp. If you came across a mite, you would have to use a magnifying glass, but you would find that it has no wings and eight legs, so it belongs with the spiders (to which it is actually related).

The pictures on page 9 show two collections of insects. The upper one is a rather mixed bag of beetles, just a few of the 250,000 different beetle species in the world. The boxes in the lower picture contain a more ordered collection of butterflies and moths grouped into families. Many people think that butterflies look better flying around in their natural surroundings, and that collecting should be reserved for thought experiments.

Other kinds of classification

Another important reason for having classification schemes is so that we can find things that are part of a large collection of similar objects. For instance, the Library of Congress in Washington, DC, houses more than 9 million books. How do you go about finding one book among so many? In fact, this library has its own classification system, and any book can be located in a matter of seconds. Most smaller libraries use the Dewey decimal system. Devised by American librarian Melvil Dewey in 1876 for the Amherst College Library, the system divides all the major subjects into 10 groups

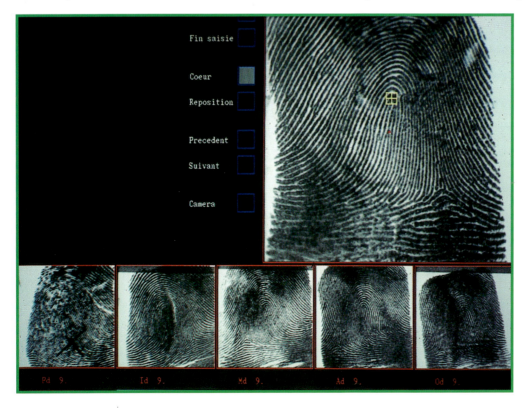

◀ **Fingerprint records** can be classified, digitized, and stored on computer, as in this French system. A fingerprint discovered at the scene of a crime can be compared with those on file, and any match can be rapidly found.

and assigns 100 numbers to each group. For example, numbers between 500 and 599 cover natural sciences and mathematics; 570 is biology, 573 is evolution, and 573.2 is human evolution.

This pattern—large groups divided into smaller groups that are themselves divided into even smaller categories, and so on—is common to nearly all scientific classification schemes. The picture on the left shows part of another enormous collection that has to be classified. Police forces throughout the word have been collecting fingerprints for over a century, ever since London's police chief Sir Edward Henry introduced the first classification scheme in 1901. Then, as now, fingerprints were sorted according to their patterns of loops and whorls in the prints. The modern FBI system, which classifies a collection of more than 90 million fingerprints, uses eight categories of loops and arches. International crime fighters are adopting computer systems to speed up the mammoth task of matching a print from a suspect or crime scene with those on record.

▲ **Lepidopterists**—scientists who study butterflies and moths—now have a complete classification scheme. Here various different groups of insects have been arranged in separate boxes.

The chemical elements

There were 63 elements known in 1869, when the Russian chemist Dmitri Mendeleev devised the classification scheme known as the Periodic Table. Since then nearly 50 more elements have been made or discovered, and they all fit into the scheme invented more than 130 years ago.

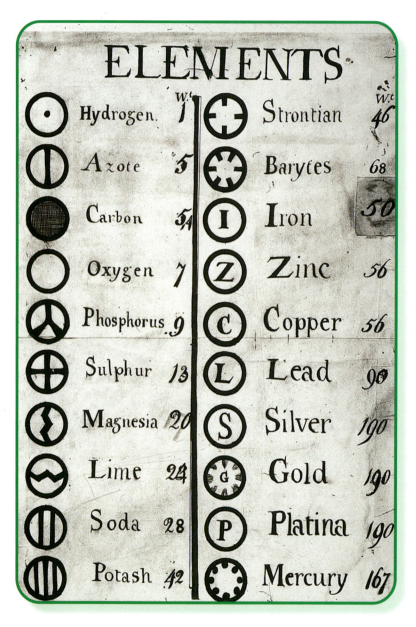

▲ **In John Dalton's list of elements** he gave each element a symbol and an atomic weight, based on taking the weight of hydrogen as 1. Some of his "elements" (such as lime) are actually compounds.

At the last count there were 112 different chemical elements, all but three of which have been given proper names. They make a mixed collection, with 99 solids, 11 gases, and two liquids at ordinary temperatures. But they all fit into Mendeleev's Periodic Table, a classification scheme that has stood the test of time for more than 130 years. A recent version of it is illustrated on pages 12 and 13.

The modern definition of an element is a substance that cannot be chemically decomposed into simpler substances. Before Mendeleev chemists had problems deciding which substances really were elements. Then, once identified, they all needed to be named. One attempt to bring order to the chaos was John Dalton's system of giving each element a symbol (see the picture on the left). Dalton was an English chemist who, in the early 1800s, measured the atomic weights (what we now call relative atomic masses) of elements, taking the weight of hydrogen as 1. When he ran out of geometrical symbols, he started using letters based on the elements' English names.

The idea of using initial letter symbols for the elements was later taken up by the Swedish chemist Jöns Berzelius. His 1828 list of atomic weights included 28 elements. Three of them— cerium, selenium, and thorium—were discovered by Berzelius himself.

Symbols instead of element names also internationalized chemistry. For example, names of copper in other European languages

► **A modern chemistry laboratory** needs ready access to many different compounds. Most labs have a storeroom containing hundreds of jars and bottles that a laboratory assistant has to arrange—usually in alphabetical order of their names—so that any compound can be found at a moment's notice. But the order would be different in Germany, say, than in America.

are *cuivre* (French), *rame* (Italian), *cobre* (Spanish), and *Kupfer* (German). But whatever they call it, chemists throughout the world recognize the symbol Cu as standing for copper. The symbol Cu, by the way, comes from *cuprum*, the Latin word for copper.

Periodic patterns

When chemists listed the elements in order of atomic weight, they saw some patterns emerging. In 1829 the German chemist Johann Döbereiner spotted groups of three elements with similar chemical properties, such as chlorine, bromine, and iodine, and calcium, strontium, and barium. Furthermore, the central element of each of these "triads," as he called them, has an atomic weight close to the average of the other two. For instance, chlorine's atomic mass is 35.5, iodine's is 126.9, and the average of these two (81.2) is close to bromine's atomic weight of 79.9.

In 1864 the English chemist John Newlands spotted another type of pattern in the list of atomic weights, which by then was rather longer.

He noted that the physical properties of the elements tended to be similar for every eighth element down the list. Thus sodium has similar properties to potassium, which was eight places further down. Despite this insight, other British chemists refused to see any significance in it.

Five years later the Russian chemist Dmitri Mendeleev was wrestling with this problem. He decided to put the names of the elements on 63 cards, one card for each element known to him. He then shuffled the cards around on a table, like playing a large game of patience. In 1869 his patience was rewarded when he homed in on an arrangement of the cards consisting of six horizontal rows (called periods) and eight vertical columns (called groups). This was the original Periodic Table, so called because the similar properties of elements appeared periodically through the table.

Sometimes an element fell into a "wrong" group. For example, iodine (atomic weight 126.9) was found in the oxygen group, below selenium, while tellurium (atomic weight 127.6) appeared

FOR MORE ON CHEMICAL ELEMENTS SEE *CHEMICAL ANALYSIS: WHAT IS IT?* 8:10; *SPECTROSCOPES* 8:20; *ROCKS AND MINERALS* 9:14

The chemical elements

The Periodic Table

The modern Periodic Table divides the chemical elements into 18 vertical groups and 7 horizontal periods. Elements in the same group have similar chemical properties. The lanthanides (with atomic numbers 57–70) are extremely similar rare metallic elements that belong between barium (Ba) and lutetium (Lu). In a similar way the actinides (atomic numbers 89–102) fit between radium (Ra) and lawrencium (Lr). What links the elements in the vertical groups is the number of electrons in the outer shells of their atoms. Having the same number means that they interact with other elements—by combining with them to make compounds, for example—in a similar way.

3 Li Lithium	4 Be Beryllium
11 Na Sodium	12 Mg Magnesium

19 K Potassium	20 Ca Calcium	21 Sc Scandium	22 Ti Titanium	23 V Vanadium	24 Cr Chromium	25 Mn Manganese
37 Rb Rubidium	38 Sr Strontium	39 Y Yttrium	40 Zr Zirconium	41 Nb Niobium	42 Mo Molybdenum	43 Tc Technetium
55 Cs Cesium	56 Ba Barium	71 Lu Lutetium	72 Hf Hafnium	73 Ta Tantalum	74 W Tungsten	75 Re Rhenium
87 Fr Francium	88 Ra Radium	103 Lr Lawrencium	104 Rf Rutherfordium	105 Db Dubnium	106 Sg Seaborgium	107 Bh Bohrium

Lanthanides

57 La Lanthanum	58 Ce Cerium	59 Pr Praseodymium	60 Nd Neodymium	61 Pm Promethium	62 Sm Samarium	63 Eu Europium

Actinides

89 Ac Actinium	90 Th Thorium	91 Pa Protactinium	92 U Uranium	93 Np Neptunium	94 Pu Plutonium	95 Am Americium

under bromine in the chlorine group. However, tellurium's chemical properties are very similar to those of selenium, while iodine's properties resemble bromine's. Mendeleev decided, quite rightly as it turned out, to swap iodine with tellurium, moving them into the columns where he felt sure they belonged.

There were many gaps in Mendeleev's original table. These "holes" corresponded, he said with great foresight, to elements that had not yet been discovered. For example, the gaps below aluminum, boron, and silicon he called eka-aluminum, eka-boron, and eka-silicon (*eka* is Sanskrit for "one," indicating that the missing element is one place lower). He even worked out the atomic weights and chief properties of the missing elements. The discoveries of gallium (in 1875 by Frenchman Paul-Emile de Boisbaudran), scandium (in 1879 by the Swede Lars Nilson), and germanium (in 1886 by the German Clemens Winkler) with just the properties Mendeleev had predicted for eka-aluminum, eka-boron, and eka-silicon showed how correct he had been. By 1914 only seven gaps remained in the table up to uranium (element 92).

Numbers not weights

Once scientists had learned about the structure of the atom, they realized that an element's chemical identity—why hydrogen is hydrogen, and helium

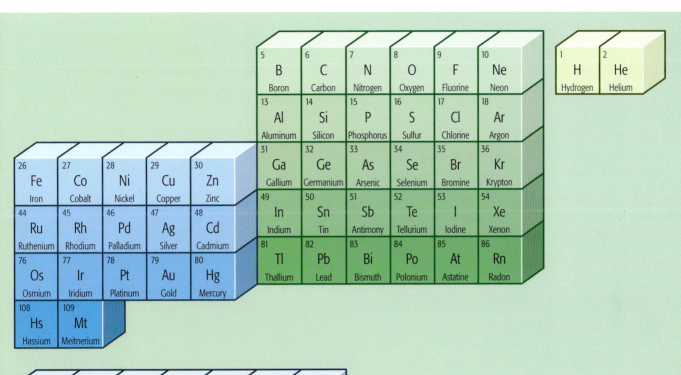

is helium—depends on the number of protons in its nucleus. This is called the atomic number (or proton number). So if we put the elements in order of atomic number, rather than atomic weight, we get the modern form of the table. The Periodic Table on these pages has the atomic numbers in the top left-hand corner of each element's box. Using atomic numbers instead of atomic weights makes tellurium and iodine, for example, fall naturally in their correct places without having to fiddle with their positions, as Mendeleev had done. Even so, his Periodic Table has been described as a most elegant solution to one of the most complex problems of classification.

Dmitri Mendeleev

Dmitri Ivanovich Mendeleev was born in 1834 in the town of Tobolsk in Siberia, the youngest of 14 children. To support the family, his mother took over the running of the local glass factory. In 1847, after the factory burned down, she took Dmitri to St. Petersburg. By 1866 Mendeleev was professor of chemistry at St. Petersburg, and three years later he announced his discovery of the Periodic Table of the elements, which was to make him the most famous Russian chemist of his time. He died in 1907. The element mendelevium (no. 101) was named in his honor.

Rocks and minerals

Minerals are naturally occurring compounds that always have the same chemical composition. Rocks are formed from minerals, and their composition can vary slightly from place to place. There are many kinds of rock and hundreds of different minerals, all of which geologists have to classify.

While theoretical chemists were working on a classification scheme for the elements, as described on the previous pages, they were too busy to worry much about where the elements came from. This was the task of the practical chemists who hunted for new elements. And the materials they usually hunted in were minerals.

Since ancient times people have been fascinated by certain minerals because of their beautiful colors and rarity. Particularly valued are sparkling gemstones such as diamond, emerald, sapphire, and ruby. When chemists analyzed these precious stones, they found that diamond is just another form of carbon (the element familiar in the forms of graphite and soot), sapphire and ruby are forms of the mineral corundum (aluminum oxide) colored by impurities, and emerald—the most valuable of all—is a complex silicate of aluminum (the mineral beryl), also colored by impurities.

They also noticed that the stones had the same chemical composition no matter where they came from. Emeralds from Colombia are chemically identical to emeralds from Australia or Zimbabwe. The same was true of other minerals. This gives

▼ **The action of wind and weather** exposes different rock layers, called strata. The rocks can be classified by type (as in the illustration on page 17) and also by age—if we assume that younger rocks always lie on top of older ones.

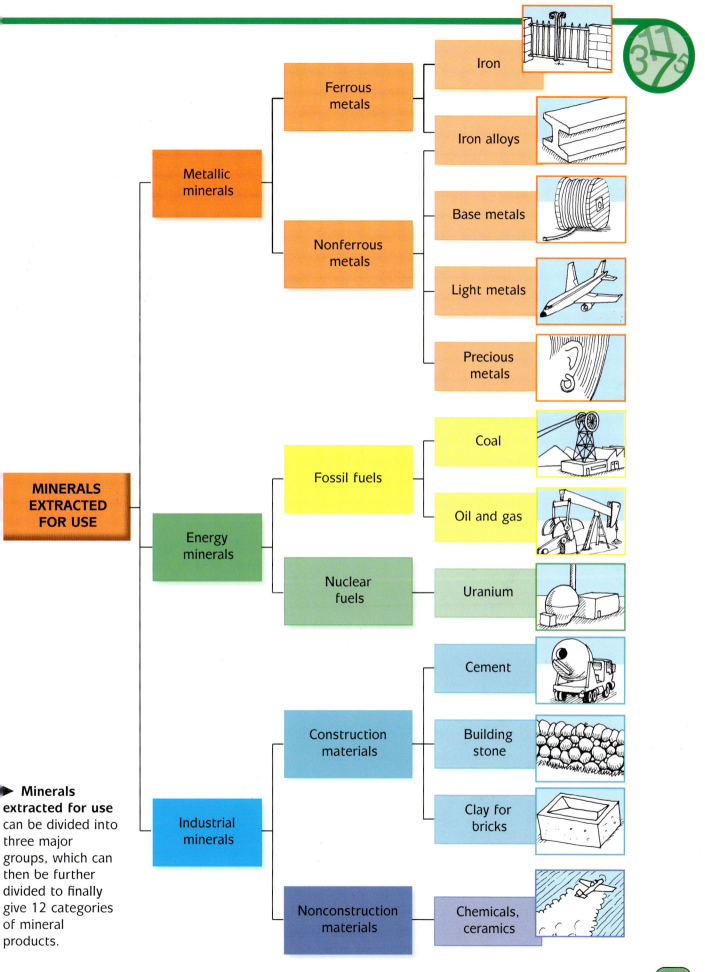

MINERALS
EXTRACTED
FOR USE

Metallic
minerals

Ferrous
metals

Iron

Iron alloys

Nonferrous
metals

Base metals

Light metals

Precious
metals

Energy
minerals

Fossil fuels

Coal

Oil and gas

Nuclear
fuels

Uranium

Industrial
minerals

Construction
materials

Cement

Building
stone

Clay for
bricks

Nonconstruction
materials

Chemicals,
ceramics

► **Minerals
extracted for use**
can be divided into
three major
groups, which can
then be further
divided to finally
give 12 categories
of mineral
products.

FOR MORE ON ROCKS AND MINERALS SEE *A LONG TIME AGO* 2:34; *CHEMICAL ANALYSIS: WHAT IS IT?* 8:10; *THE CHEMICAL ELEMENTS* 9:10

Rocks and minerals

us our definition of a mineral: a naturally occurring chemical compound that has the same composition no matter where it is found. There are also a few elements, such as gold, that are found naturally as minerals, and they obviously are always chemically the same.

Classifying minerals

Today more than 3,500 minerals have been identified and named. Another challenge for classification! One way to classify minerals is by their chemical composition. They can be allocated to groups such as oxides, sulfides, sulfates, phosphates, silicates (the largest group), and so on. Because all but two or three minerals are crystalline solids, they can also be classified by their crystal form. Crystallographers distinguish between *crystal habit*, which is the shape of the crystal, and *crystal structure*, which depends on how atoms are arranged in the lattice that makes up the crystal's internal network.

Other classification schemes are based around the physical properties of minerals. Color and luster (sheen) have been used, but a more testable property is hardness. Devised by the German mineralogist Friedrich Mohs in 1812, the Mohs scale of hardness is a list of ten minerals, each of which is hard enough to scratch the surface of all the minerals above it in the list (see below).

THE MOHS HARDNESS SCALE FOR MINERALS

Mineral	Hardness	Mineral	Hardness
Talc	1	Orthoclase	6
Gypsum	2	Quartz	7
Calcite	3	Topaz	8
Fluorite	4	Corundum	9
Apatite	5	Diamond	10

▼ **Rocks and minerals** come in a wide range of colors and textures. This collection includes specimens from all over the world, which the collector has arranged in an attractive display.

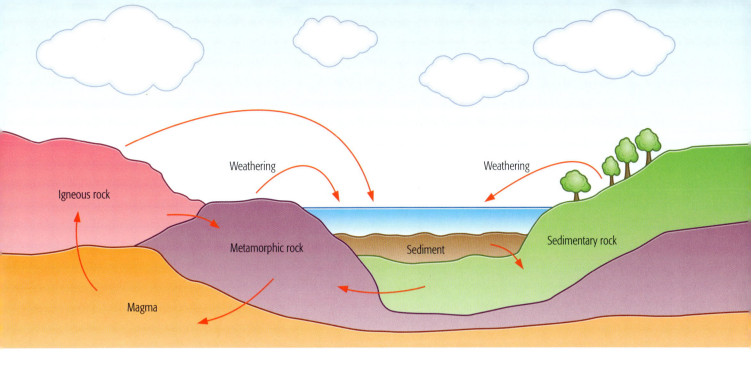

Igneous rock

Weathering

Weathering

Metamorphic rock

Sediment

Sedimentary rock

Magma

Diamond is the hardest known natural substance and will scratch any other mineral. That is why you can test whether or not a diamond is genuine by seeing whether it will scratch glass (hardness up to 6). Here are some other hardness values: fingernail, 2.5; copper coin, 3; knife blade, 5 to 6; steel file, 6 to 7. Geologists trying to identify a mineral they come across in the field often scratch it to get an idea of its hardness with reference to these values.

Many minerals are extracted from the ground for use as fuels or as raw materials for industry. Minerals are converted to a huge range of useful materials, from metals for automobiles to chemicals for fertilizers. A breakdown of uses of minerals is given on page 15 (this is yet another kind of classification scheme).

The rocks go around and around

Minerals combine with one another to form rocks. For example, the minerals amphibole, feldspar, mica, and quartz are the components of granite. But their proportions vary from place to place—not every piece of granite is the same as every other piece (that is what makes rocks different from minerals).

Geologists classify rocks into three main types. *Igneous rocks* originate in the molten magma beneath the Earth's crust. They rise to the surface

▲ **The origin of igneous rock** is molten magma below the ground, while sedimentary rock forms from deposits produced by the weathering of all kinds of rocks. High temperatures and pressures below ground convert either igneous or sedimentary types into metamorphic rock. These types of rocks are continually changing into one another in a sequence called the rock cycle.

and solidify above the ground or sometimes burst forth from volcanoes. Igneous rocks are among the hardest of all rocks.

The action of wind and rain on igneous rocks—the process called weathering—gradually breaks them down into fine grains, such as sand. Grains that settle at the bottom of the sea and accumulate as a sediment can become compressed to form *sedimentary rocks*. Sand grains, for example, form sandstone. Calcium salts from the skeletons of countless small sea creatures also settle and form sedimentary rocks such as limestone.

Beneath the ground heat and pressure get to work on sedimentary rocks (and some igneous ones) and change them into what are called *metamorphic rocks*. Marble is a metamorphic rock, formed from limestone. The new rocks also become weathered, and the whole process goes around again in what is called the rock cycle (see the illustration on this page).

Classifying plants

In 1735 the Swedish botanist Carolus Linnaeus made the first attempt to classify living things. He classified plants and invented a system for naming them in which each plant was given two names—a genus and a species. This system is still in use today.

One of the products of the gradual weathering of rocks is soil. And soil provides the nurturing environment needed by plants. A scientific definition of a plant is a many-celled organism that gets its body chemicals through photosynthesis. This is the process that works in green leaves, using the Sun's energy to convert water (from the soil) and carbon dioxide (from the air) into chemicals needed by the plant, such as starch and sugar. Plants cannot move around—they are literally rooted to the spot; they have no nerves or senses, and the walls of their cells are made of cellulose.

Essentially, that's all a plant is. But an amazing number of variations on this basic theme have evolved on Earth over millions of years. There are tiny mosses, colorful flowers, and enormous trees. Plants live at the bottom of the sea and at the tops of the highest mountains, in dry deserts and in permanently waterlogged swamps. So how on Earth can we classify such a huge variety of organisms?

Sorting by shape

One method, like many other classification schemes, relies on appearance. You could, for example, divide most plants into trees, shrubs, flowers and herbs, and grasses. But that would not get us very far. When does a shrub become a tree—how tall does it have to be?

▼ **In collections of plants** you can see similarities and differences between plants from similar habitats. These desert plants include cacti, succulents, and various kinds of palms.

And is tall bamboo a shrub? (No, it's actually a type of grass.) We can, however, name and classify *parts* of plants by their appearance.

The pictures on this page show how it can be done for leaves. The larger picture illustrates different leaf shapes with their botanical descriptions. But not only does the overall shape vary, so does the shape of the edges (called leaf margins). Eight different types are shown in the smaller illustration. The scientific names for the shapes owe a lot to Latin, commonly used in science at the time of the early botanists.

A different approach is illustrated on page 20, which shows the names used in the classification of fruit. Notice that, botanically speaking, any plant part that bears a seed or seeds is a fruit: Nuts, peas, and the capsules on poppies are all kinds of fruits.

Leaf margins

Crenate (scalloped)
Dentate (toothed)
Entire (complete)
Incised (cut in)
Serrate (saw-toothed)
Sinuate (winding)
Spinose (spiny)
Undulate (wavy)

▲ **The shapes of the edges of leaves**, called leaf margins by botanists, provide one way of classifying leaves. Eight different types are illustrated here.

Leaf shapes

Cordate (heart-shaped)
Hastate (spear-shaped)
Spatulate (like a spatula)
Lanceolate (lance-shaped)
Sagittate (arrow-shaped)
Peltate (like a shield)
Linear (like a line)
Palmate (palm-shaped)
Ovate (egg-shaped)
Orbicular (round)

◄ **An additional way** of classifying leaves is by their overall shape, each of which has a special botanical name.

FOR MORE ON CLASSIFYING PLANTS SEE *COMPOUND MICROSCOPE* **5:16;** *CLASSIFYING ANIMALS* **9:24;** *CLASSIFYING BACTERIA* **9:32**

Classifying plants

A berry, like a tomato or a grape, is a fleshy fruit containing seeds. It does not break open to release its seeds, but falls to the ground and rots. Many berries are brightly colored and provide food for birds, and so can be spread far and wide. The outer skin may be tough, as in a hesperidium (like the orange and other citrus fruits) and in a pepo (like melons and squashes).

A drupe is a fleshy fruit with a central seed (or seeds) that has a thin hard "skin," as in a cherry pit. Other drupes are peaches, plums, and holly berries (which, botanically, are not true berries).

Another common type of fruit is a pome, as typified by apples and pears. The real fruit is at the core, surrounding the seeds, and is itself surrounded by the fleshy part that we eat. The seeds of legumes grow as pods usually containing several seeds. As the pod dries, it splits open to release the seeds. Twisting of the pod may aid this action; in some plants the pod bursts with explosive force to scatter the seeds. Beans and peas are common legumes that are cultivated as vegetables.

The maple has a different way of spreading its seeds. The seeds are paired to form a winged structure called a double samara, which spins like a helicopter as it falls from the trees. An ordinary samara is a single seed with one wing.

Sorting by function

The systems just described are fine for plants with prominent leaves and fleshy fruits. But what about grasses, for example, which all look much the same at first sight? A more promising approach is to classify plants in terms of the way they work. To do that, botanists divide the 260,000 or so known plants into ten major groups called divisions, which all have names ending in -*phyta*.

The largest group is the Magnoliophyta, which include most plants that reproduce using seeds that are enclosed within a fruit. Nine out of every ten plants have enclosed seeds, and such plants are called angiosperms. Angiosperms include garden flowers and wild flowers, herbs, trees, shrubs, and grasses. They

Types of fruit

Berry — Tomato
Capsule — Poppy
Double samara — Maple
Drupe — Peach
Hesperidium — Orange
Legume (pod) — Pea
Nut — Hazel
Pepo — Squash
Pome — Apple

◄ **There are various types of fruit**, each with a special name. Nine of the common types are illustrated here. Some fruits that we call berries, like blackberries, are in fact aggregate fruits— a blackberry consists of many drupelets.

provide us with food (fruits, roots, and grains) and materials for construction (timber and bamboo) and to make paper and cloth (wood pulp and cotton).

When an angiosperm seed germinates, the first thing to appear out of the ground is a seed leaf or two. The botanical name for a seed leaf is cotyledon. Plants with one seed leaf, like grasses, lilies, and most spring bulbs, are called monocotyledons. Plants with two seeds leaves, which includes most other plants, are called dicotyledons.

The group called gymnosperms bear naked seeds, usually borne in cones. They do not produce flowers. The gymnosperms include trees with

These flowers, all kinds of daisies, belong to the family Compositae, the largest of all plant families. But each belongs to a different genus and is a unique species: black-eyed Susan (*Rudbeckia hirta*, above), ox-eye daisy (*Chrysanthemum leucanthum*, left), and dandelion (*Taraxacum officinale*, below).

Classifying plants

Kingdom	Plantae	All plants
Division	Anthophyta	Flowering plants
Class	Dicotyledonae	Dicotyledons
Order	Fagales	Beeches
Family	Betulaceae	All birches
Genus	*Betula*	Northern birches
Species	*pendula*	Silver birch

▶ **All plants** belong to the kingdom Plantae, which is subdivided into divisions, classes, orders, families, genera, and finally species. Shown above is the complete classification of the common silver birch (right).

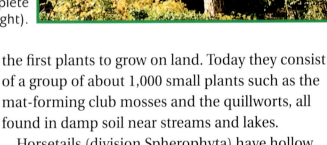

needle-shaped or scaly leaves, such as cedars, firs, pines, and spruces, most of which are evergreens. They form the division of up to 575 species called Coniferophyta. Also included among gymnosperms are more primitive ancient trees such as the 120 species of palmlike cycads (division Cycadophyta) and the ginkgo. There is now only one species of ginkgo, placed in its own division, Ginkgophyta.

The next division is Filicophyta, the 12,000 or so species of ferns. They are easy to recognize by their feathery leaves, called fronds. They grow from underground stems and first appear as tightly coiled structures that in some places are known as fiddleheads because they resemble the scroll on a violin. They vary in size from small aquatic plants to giant tree ferns up to 20 meters (65 ft) tall. They thrive best in damp woods and bear their "seeds" (actually spores) on the undersides of the fronds.

Lycophytes (division Lycophyta) were among the first plants to grow on land. Today they consist of a group of about 1,000 small plants such as the mat-forming club mosses and the quillworts, all found in damp soil near streams and lakes.

Horsetails (division Spherophyta) have hollow jointed stems with no proper leaves. Modern types grow no more than a meter (nearly 3 ft) tall, but their ancient ancestors were once the largest plants on land. The stem tissues contain tiny grains of silica (sand), which explains the plant's old name of scouring rush—people once used horsetails to clean their pots and pans.

Mosses, liverworts, and hornworts make up the eighth division, Bryophyta. There are 16,000 species of bryophytes. They have no true roots and cling to the soil with hairlike filaments called rhizoids. Bryophytes are among the most widespread of plants, living in moist places throughout the world.

The last two divisions are the unusual Psilophyta (or whisk ferns) and the Gnetophyta, a group of about 70 cone-bearing plants of desert regions. Algae (pondweeds and seaweeds) and fungi (mushrooms) are not part of this scheme. They are not considered to be true plants and are given kingdoms of their own.

The naming system

Some of the plant divisions are split into classes. For example, the Magnoliophyta form the classes Dicotyledonae (dicotyledons) and Monocotyledonae (monocotyledons). Classes are broken down further into orders, and orders into families. Examples include the order Fagales, which includes all the beech trees, and the family Compositae, to which all kinds of daisies belong. The final two levels of classification—genus and species—identify a specific plant.

A species is a plant that can breed only with another one of its kind. A black-eyed Susan is a kind of daisy, as is an ox-eye daisy. But they cannot interbreed, and so they are different species.

A genus is a group of closely related species. A plant's formal, or taxonomic, name has two parts: the genus name and the species name. The two daisies just mentioned are *Rudbeckia hirta* and *Chrysanthemum leucanthum*. Notice that these names, all of which are based on Latin, are printed in italic letters, with the genus (only) beginning with a capital letter.

This two-name system, formally known as binomial nomenclature, was thought up in the mid-1700s by Swedish botanist Carolus Linnaeus. It has one tremendous advantage: A plant's taxonomic name is the same the world over, even if—as is usually the case—the common name for the plant varies from place to place and from one language to another.

Carolus Linnaeus

Karl von Linné (Carolus Linnaeus is the Latin version of his name) was a Swedish botanist who was born in 1707 in Råshult. His father was a pastor. In 1727 he went to the University of Lund to study medicine, but a year later he moved to Uppsala and switched to botany. By 1730 Linnaeus had worked out his system of classifying plants, first announced in 1735 in his book Systema naturae *("The System of Nature"), which allocated all things to one of three kingdoms: animals, plants, and minerals. In 1749 he introduced binomial nomenclature, the two-level system for naming plants according to their genus and species. After his death in 1778 the system was extended to animals.*

Classifying animals

The two-name system that Linnaeus devised for plants is also used for animals. The classification and naming of living things is a science of its own, called taxonomy. It reveals relationships between similar organisms and sometimes provides clues about their evolution.

Animals are classified in the same way as plants. The only difference is that what is called a division in plants is called a phylum (plural phyla) in animals. Together, the 1.5 million known species of animals—three-quarters of the species of living things on Earth—make up the kingdom Animalia.

Like plants, animals are made up of many cells. But animals differ from plants in several important ways. Most animals can move around to look for food or to escape from their enemies. To do so they need muscles, controlled by a brain and a nervous system. Senses provide the brain with information about the exterior world. And unlike a plant, an animal cannot use simple chemicals as "food" to build up its tissues. It has to eat other organisms, either plants or other animals.

Animals live throughout the world: in deserts, in the seas, and on the tops of mountains. They walk, crawl, swim, and fly, and parasites can even live inside other animals. The largest living animal is the blue whale, which has been known to grow up to 33 meters (108 ft) long. But most animals are less than 2.5 centimeters (1 in.) long—because most of them are insects.

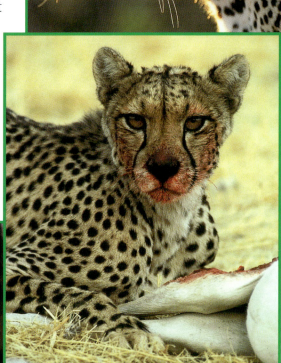

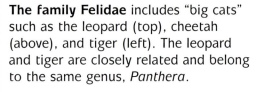

The family Felidae includes "big cats" such as the leopard (top), cheetah (above), and tiger (left). The leopard and tiger are closely related and belong to the same genus, *Panthera*.

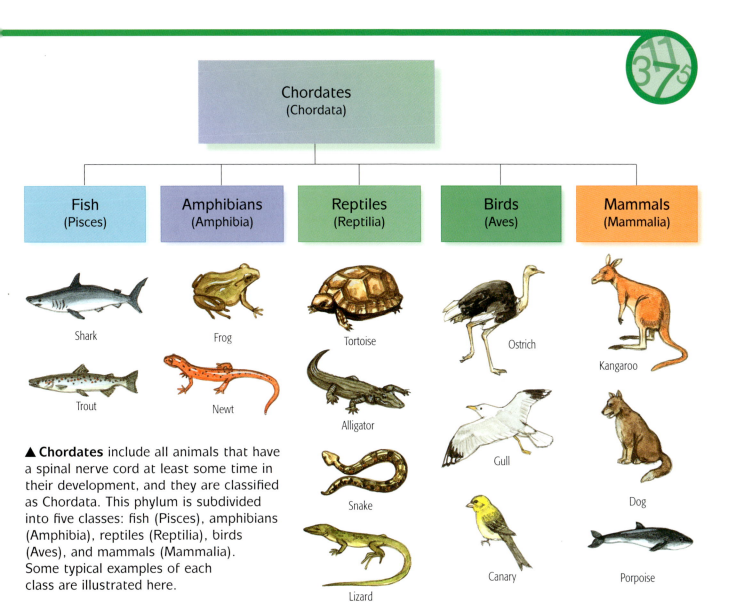

| Chordates (Chordata) |
| Fish (Pisces) | Amphibians (Amphibia) | Reptiles (Reptilia) | Birds (Aves) | Mammals (Mammalia) |

Shark

Trout

Frog

Newt

Tortoise

Alligator

Snake

Lizard

Ostrich

Gull

Canary

Kangaroo

Dog

Porpoise

▲ **Chordates** include all animals that have a spinal nerve cord at least some time in their development, and they are classified as Chordata. This phylum is subdivided into five classes: fish (Pisces), amphibians (Amphibia), reptiles (Reptilia), birds (Aves), and mammals (Mammalia). Some typical examples of each class are illustrated here.

Modern animal classification

The animal kingdom is divided into 14 main phyla. All but one of them are invertebrates, which means animals without backbones. They consist of various kinds of worms, mollusks, and arthropods—animals with jointed legs such as arachnids (e.g., spiders), crustaceans (e.g., crabs and barnacles), and insects. The 14th phylum is the chordates (phylum Chordata), which are animals that have a spinal cord at some time in their development. Practically all of the adult chordates are animals with backbones.

The next taxonomic level down from phylum is class, and there are five classes of chordates (see the diagram on this page). They are, with Latin names in brackets, fish (Pisces), amphibians (Amphibia), reptiles (Reptilia), birds (Aves), and mammals (Mammalia).

Classes are split into orders. For example, there are four orders of reptiles: turtles and tortoises (Chelonia), snakes and lizards (Squamata), alligators and crocodiles (Crocodilia), and the tuatara (Rhynchocephalia). The tuatara is a one-off, a lizardlike animal that lives in New Zealand and is the sole survivor of a group whose other members have been extinct for 100 million years.

The mammals split into three groups: the egg-laying monotremes, such as the Australian echidna (spiny anteater); the marsupials, such as the opossums and kangaroos; and the placental mammals—all the rest. As you can see from the table on page 26, there are 17 orders of placentals, ranging

FOR MORE ON CLASSIFYING ANIMALS SEE *GENETIC FINGERPRINTING* 8:28; *CLASSIFYING PLANTS* 9:18; *CLASSIFYING BACTERIA* 9:32

Classifying animals

from bats and carnivores to seals and primates (the order to which humans belong). Orders are further split into families, families into genera, and genera into species. The complete classification of the African lion is given as an example on page 27.

What belongs where?

How do zoologists decide how to classify mammals? For instance, dogs and cats are placed with the meat-eating carnivores. But dolphins and seals eat meat (fish), and so do some monkeys, but they are not in with the carnivores. As you can see from some of the descriptions in the table below, such as "even-toed" and "toothless," the structure of the mammal can provides a basis for classification. One such approach classifies mammals—or at least allocates them to one of the existing categories—according to their teeth.

Mammals have up to four different kinds of teeth (see the illustration on page 27), called incisors,

canines, premolars, and molars. Incisors sit at the front of the jaw and are used for biting off pieces of food. Listen to a cow grazing, and you will actually hear the incisors biting off the grass. Well-developed incisors are therefore often a sign that the animal is a herbivore, a plant-eater, probably belonging to one of the orders of hoofed mammals.

Canines are sharp stabbing teeth used for grasping and holding onto food, which is often alive and struggling when it is first grasped. These fanglike teeth reached their greatest development in the extinct saber-toothed cats, which had canines long enough to penetrate the thick wool-covered skin of mammoths. All cats have long canines—look at the pictures of the tiger and the lion on pages 24 and 27. Dogs and hyenas also have long canines.

The premolars chop food into chunks, either ready for swallowing or for further chewing. In carnivores the premolars are called carnassials.

ORDER	DESCRIPTION	EXAMPLES
Artiodactyla	Even-toed hoofed animals	Antelopes, deer, pigs, hippos
Carnivora	Meat-eating animals	Cats, dogs, bears, skunks
Cetacea	Whales, dolphins, porpoises	Humpback whale, rorqual
Chiroptera	Bats	Fruit-eating bats, vampire bats
Dermoptera	Colugos	Flying lemurs
Edentata	Toothless mammals	Anteaters, armadillos, sloths
Hyracoidea	Hyraxes	Rock hyrax
Insectivora	Insect-eating mammals	Shrews, moles, hedgehogs
Lagomorpha	Rabbits, hares, pikas	Cottontails, brown hare
Perissodactyla	Odd-toed hoofed animals	Horses, rhinoceroses, tapirs
Pholidota	Pangolins	Cape pangolin
Pinnipeda	Seals	Walrus, sea lions, elephant seal
Primates	Apes, monkeys, humans	Gorillas, baboons, humans
Proboscidea	Elephants	African elephant, Asian elephant
Rodentia	Rodents	Rats, mice, squirrels, porcupines
Sirenia	Sea cows	Dugong, manatee
Tubulidentata	Aardvark	Aardvark (only 1 species)

▶ **All animals** belong to the kingdom Animalia, which is subdivided into phyla, classes, orders, families, genera, and finally species. On the right is the complete classification of the African lion.

◀ **Placental mammals** are classified in 17 different orders, here listed alphabetically. There is a wide variation in numbers of species in each order, from only one—the aardvark—in the order Tubilidentata to more than 1,800 species of rodents (in the order Rodentia).

They are shaped like the blades of shears, and they slice off food when the animal bites with the side of its mouth. The molars, behind the premolars, grind the food into small pieces.

KINGDOM	ANIMALIA	ALL ANIMALS
Phylum	Chordata	Animals with a spinal cord
Class	Mammalia	All mammals
Order	Carnivora	Flesh-eaters
Family	Felidae	All cats
Genus	*Panthera*	Big cats
Species	*leo*	Lion

▶ **From a mammal's teeth** we can tell its diet and therefore the major group it belongs to. Herbivores, for example, have large molars to grind up their tough plant food, while meat-eating carnivores have scissorlike premolars to slice up their animal prey.

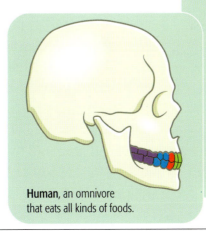

Human, an omnivore that eats all kinds of foods.

Mammal teeth

Mammals have four types of teeth:
Incisors (green) bite off pieces of food.
Canines (red) grasp and hold food.
Premolars (blue) chop the food into chunks.
Molars (purple) grind it into small pieces.

🟩	Incisors
🟥	Canines
🟦	Premolars
🟪	Molars

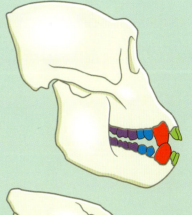

Chimpanzee, a climbing herbivore that eats leaves.

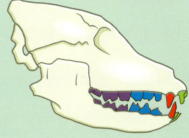

Wolf, a carnivore that eats other animals.

Rat, a rodent that gnaws mainly hard foods.

Shrew, an insectivore that feeds mainly on insects.

Horse, a herbivore that grazes on grass.

Classifying animals

▲ **Five bee-eaters** rest on a branch during their annual migration from Australia to Indonesia. Their Latin name is *Merops ornatus*, which serves to distinguish them from seven other kinds of bee-eaters.

The skulls in the illustration on page 27 are not drawn to scale; they are drawn to the same size so that you can better compare the different patterns of teeth. The skull at the bottom left shows that we humans possess all four kinds of teeth. That suits us well because we are omnivores and can eat any kind of food (though some people choose not to).

Other major groups

Allocating mammals to the correct phylum is fairly easy. As somebody once said, "Mammals are the furry ones with a leg at each corner"—usually true, unless it's a whale! But some phyla contain what appears to be an odd assortment of animals.

Consider the mollusks. They are a group of unsegmented invertebrates and in this respect are unlike the various phyla of worms, which have bodies divided into segments. Most are slow-moving creatures with shells. They include the shield-shaped chitons and the long tube-shaped tusk shells. There are nearly 1,500 species in these two classes, and they all live at the bottom of the sea.

Bivalves—the mollusk class containing shellfish with a pair of hinged shells, such as cockles, mussels, and oysters—total 8,000 different species. Even this number looks small when compared with the gastropods. This class of mollusks (their name means "stomach foot") includes 35,000 species of slugs, snails, and limpets. They can move around better than most mollusks; and if they have a shell, they carry it around with them. Slugs and snails are the only mollusks that live on land.

But the fastest and most intelligent mollusks of all are the cuttlefish, octopuses, and squids. Except for the nautilus, they have no shell, but they do share another useful feature—tentacles. An octopus, as its name suggests, has eight tentacles, while a squid has ten. A giant squid can have tentacles up to 15 meters (50 ft) in length, and its long snakelike "arms" are thought to be the origin of old sailors' tales about sea serpents.

The biggest phylum

Mollusks may include the most varied creatures in a single phylum, but for sheer numbers they are left far behind by the arthropods, with well over a million species at the last count. Their name means "jointed legs," a characteristic they nearly all share. They were the first animals to conquer the land and soon also took to the air. Today they are found in every habitat on Earth. There are 13 classes of arthropods, although some classes contain only a few hundred species.

There are at least 30,000 species of crustaceans, with new ones being discovered all the time. This arthropod class includes crabs, lobsters, prawns, shrimps, and pill bugs. Barnacles, which attach themselves permanently to rocks like a limpet, also belong to the crustacean class.

Most crustaceans have many legs—ten is a popular number. Not far behind, in terms of legs, are the arachnids, which all have eight legs. This class includes scorpions, spiders, daddy longlegs (harvestmen), and mites and ticks. Together they total more than 50,000 species. Some are poisonous, and nearly all of them bite! There are more than 35,000 species of spiders, for example, ranging in size from under 0.1 millimeters (0.03 in.) to the huge 25-centimeter (10-in.) bird-eating spider.

Insects have to get by on just six legs, but the class makes up for this by having more than a million species divided into 29 different orders. Some of the better-known orders are ants, bees, and wasps; beetles; bugs; butterflies and moths; dragonflies; earwigs; flies; fleas; grasshoppers and crickets; lice; roaches; termites; and walkingsticks. We tend to call any insect a "bug," but in classification terms bugs are a single order of insects that suck blood or plant sap.

▼ **A shoal of blue-green chromis fish** dodges in and out of a coral reef. Their Latin name *Chromis viridis* means, not surprisingly, blue-green chromis!

Blood groups

Most early attempts at making blood transfusions between people ended in failure because the body reacted against the "foreign" blood and rejected it. Once the American pathologist Karl Landsteiner had discovered blood groups and showed that only some groups were compatible with others, successful transfusions became routine.

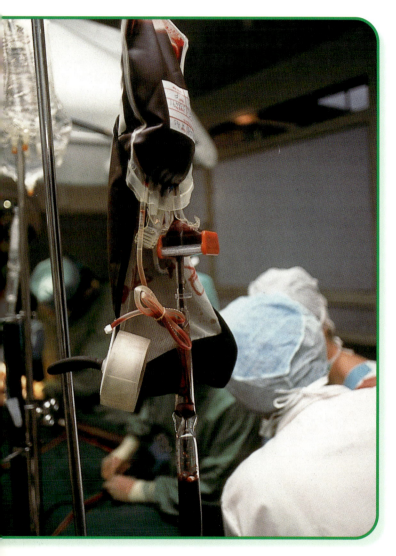

A man of average height and weight has about 5 liters ($10\frac{1}{2}$ pints) of blood coursing through his arteries and veins. A woman has a pint less. Blood is red because it is made up mainly of red blood cells—about 25 billion of them in an adult's body. They are manufactured mainly in the bone marrow and have a life of about 110 days.

If the used-up red cells are not replaced fast enough, the result is the disorder anemia. In the 19th century many people became ill or died because of anemia. Others died because they lost blood through injury or during childbirth. Early attempts to replace lost blood with a transfusion of blood from other people or from animals nearly always ended in failure.

▲ **A patient undergoing surgery** receives a life-saving blood transfusion. The blood must be correctly matched to the patient's blood group.

▶ **A pint bag of whole blood**, labeled Group B, Rhesus-negative, is placed in a refrigerated blood bank and kept between 1 and 6°C for up to 3 weeks.

Different kinds of blood

We now known that these human-to-human transfusions failed because the bloods were of different types, or groups. Blood groups were discovered in 1901 by the Austrian-American Karl Landsteiner. He found that blood contains substances called antibodies and antigens. When an antigen enters a person's blood, it stimulates the blood to produce antibodies against it. Landsteiner recognized three groups of antigens, which he called A, B, and O. Sometimes these substances caused the red cells of transfused blood to clump together when it mixed with blood of another group. Other combinations did not cause clumping, and the transfusion was successful. A year later a fourth group, called AB, was also recognized.

The chart at the top of this page shows which groups can safely be mixed. For example, group A will obviously mix with A, and it is also safe with group AB. Group B will mix with groups B and AB, while group AB will mix only with another batch of AB blood. Group O, which lacks any kinds of antigens, will happily mix with any other group: A, B, AB, and, of course, O. Because it will mix with any other group, it is called the universal donor. A person with AB blood can be transfused with any other group and is therefore called the universal recipient.

Help from monkeys

In the 1940s experimental work on immunization often used blood cells from species of rhesus monkey. In 1940 Landsteiner and his coworker Alexander Wiener discovered another antigen in human blood, which they called the rhesus (Rh) factor. People who have it in their blood are called rhesus positive (Rh+), and the far less numerous people lacking it are called rhesus negative (Rh−).

Blood donor	Recipient			
	A	B	AB	O
A	✓	✗	✓	✗
B	✗	✓	✓	✗
AB	✗	✗	✓	✗
O	✓	✓	✓	✓

✓ = Successful
✗ = Unsuccessful

▲ **The results of blood transfusions** between all possible combinations of blood groups indicate that anyone can receive type O blood (the universal donor) and a person with type AB blood (the universal recipient) can receive a transfusion of any other type.

Karl Landsteiner

Born in Vienna in 1868, Karl Landsteiner studied medicine at the university there, and after graduating in 1891, he studied chemistry at various European universities. In 1922 he went to work at the Rockefeller Institute in New York and later became an American citizen. Landsteiner is best remembered for discovering the ABO system of blood groups, which he announced in 1901. In 1908 he discovered that poliomyelitis is caused by a virus. For his work on blood groups he was awarded a Nobel Prize in 1930. In 1940, when he was over 70 years old, he discovered the rhesus factor in human blood. He died three years later.

FOR MORE ON BLOOD GROUPS SEE *COMPOUND MICROSCOPE* **5**:16; *GENETIC FINGERPRINTING* **8**:28; *CLASSIFYING BACTERIA* **9**:32

Classifying bacteria

Bacteria make up one-half of one of the five kingdoms into which all living things are classified. Originally they were classified by their shapes, according to whether they were spheres, oblongs, or spirals. A later classification method is based on how they react to certain dyes.

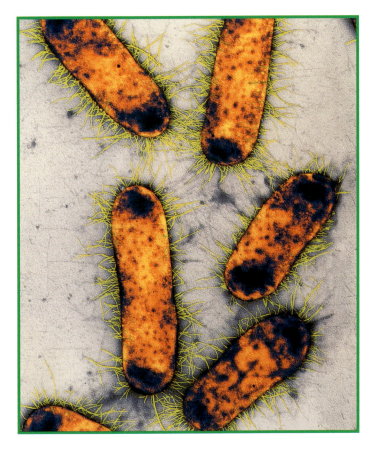

▲ **In the five-kingdom scheme** of classification, bacteria (above) belong to the kingdom Monera. They form their own phylum (below) and share their kingdom with the phylum Cyanophyta (the blue-green algae).

Bacteria are microscopic single-celled organisms that lack a nucleus in their cells. Most are harmless, though a few cause diseases in humans. The harmful types release poisonous substances called toxins into the bloodstream of the person they infect. Severe bacterial diseases include cholera, diphtheria, leprosy, tetanus (lockjaw), tuberculosis, and the type of food poisoning called botulism. In the five-kingdom system of classification bacteria make up one of the two phyla in the kingdom Monera (monerans). There are about 1,600 species, but detailed classification has proved difficult.

Sorting by shape

Bacteria (singular bacterium) range in size from 1 micrometer (1/25,000 in.) to ten times as big. They can be seen only through a microscope. Because some cause such serious diseases, they have been intensely studied over the years. One thing that becomes immediately apparent when looking at bacteria is that they differ widely in shape. For this reason shape was the first way of classifying them.

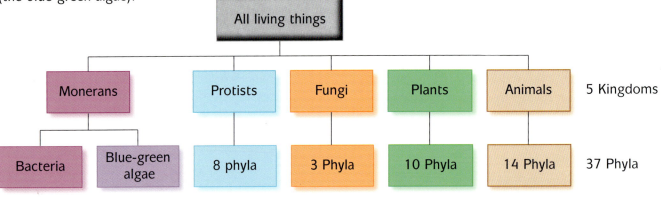

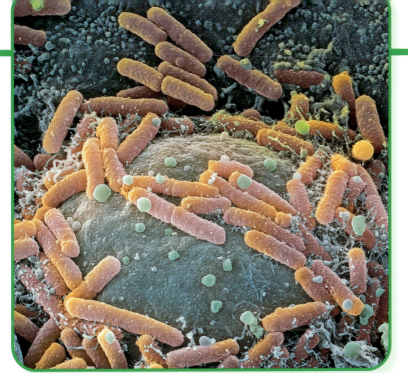

◀A colored electron micrograph shows a group of bacteria called *Escherichia coli*, which live in small numbers quite normally in the human intestine. Their rodlike shape confirms that they are a type of bacillus. A few near the center of the photograph are dividing in two—their way of reproducing.

Two common bacterial shapes are spheres and rods—the word bacterium comes from the Greek word *bakterion*, which means "little staff." The spherical type is called a coccus (plural cocci). It may exist singly (monococcus), in pairs (diplococcus), in chains (streptococcus), or in small groups resembling bunches of grapes (staphylococcus). They are illustrated on the right.

A rod-shaped bacterium is called a bacillus (plural bacilli). Bacilli may be short and fat or long and slender. Most are smooth, but some sprout many fine hairs (called flagella), which they beat like oars to move around in their fluid environment. All types reproduce by simply splitting in two. You can see some bacilli dividing near the center of the picture above.

Other bacteria shapes are based on spirals. Corkscrew-shaped spirochetes can move rapidly. One type of spirochete causes the diseases syphilis and yaws. Spirilla are also fast movers, with their tufts of flagella at each end. Most feed on dead organic matter, but one species causes rat-bite fever in humans. Vibrio is the name of a comma-shaped genus of bacteria that live in moist soil and water. Some live as parasites in humans, including the bacterium that causes cholera.

Another way of classifying bacteria into two groups depends on how they change color when treated with a special dye called Gram's stain, invented in 1884 by Danish bacteriologist Hans Gram. Gram-positive bacteria stain purple, while Gram-negative ones stain pale pink.

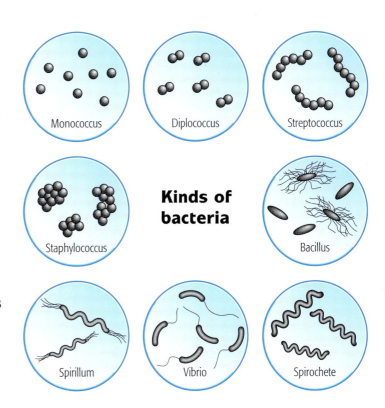

Kinds of bacteria

Monococcus

Diplococcus

Streptococcus

Staphylococcus

Bacillus

Spirillum

Vibrio

Spirochete

▲ **Bacteria can be classified** by their shapes. The spherical type called a coccus (plural *cocci*) is one of the most common. The rod-shaped bacillus (plural *bacilli*) may have hairlike appendages. All of these types are capable of causing disease.

FOR MORE ON CLASSIFYING BACTERIA SEE *COMPOUND MICROSCOPE* **5**:*16*; *ELECTRON MICROSCOPES* **6**:*20*; *CLASSIFYING PLANTS* **9**:*18*

Organizing the heavens

Objects we can see in the night sky range from planets and their moons to stars and the vast collections of stars called galaxies. Astronomers have devised classification schemes for most of them, which also give clues about how the objects were formed.

When you look up at the sky on a cloudless night, you can see many points of light. Most of them are stars, but there are also planets, and you might even see a comet or a meteor. Together the planets, stars, galaxies, and all the other heavenly bodies make up the universe.

Stars and constellations

By far the most numerous objects in the night sky are stars. Our Sun is a typical star, a huge mass of

The largest objects in the heavens are galaxies, which are made up of thousands of stars. Pictured here are the constellation Orion (above) and the spiral galaxy of Andromeda (above right). The bright object pictured on the right is a nebula around the star called Xi Cygni.

luminous gas that produces energy by nuclear fusion reactions in its core. In the most common reaction hydrogen atoms join to form helium atoms. There are various kinds of stars, depending mainly on their mass. Some, called binary stars, occur in pairs, swinging constantly around each other like waltzing dancers. Others are varying in brightness, expanding or shrinking, being created out of dust and gas, or dying in the spectacular explosion of a supernova.

Many ancient peoples saw patterns in the stars. The Greeks and Romans gave names to various groups, called constellations, which we still recognize today. There are 88 constellations, with names like Ares (the Ram), Draco (the Dragon), Leo (the Lion), Pegasus (the Winged Horse), and Taurus (the Bull). Few of the stars in any constellation are associated with one another. Most are not even close together—they lie at different distances and just appear to be close when viewed from Earth.

Giants and dwarfs

Ejnar Hertzsprung was a Danish astronomer who realized (in about 1905) that there is a connection between a star's brightness and its color or temperature. The brighter a star is, the more blue is its color and the hotter it is. In 1913 the American astronomer Henry Russell discovered the same relationship. He drew a chart to illustrate it, now known as the Hertzsprung–Russell diagram (as illustrated below).

Stars that follow the brightness–temperature relationship form a broad band called the main

▼ **The Hertzsprung–Russell diagram** classifies stars in terms of their brightness and surface temperature. The Sun lies on a diagonal band called the main sequence.

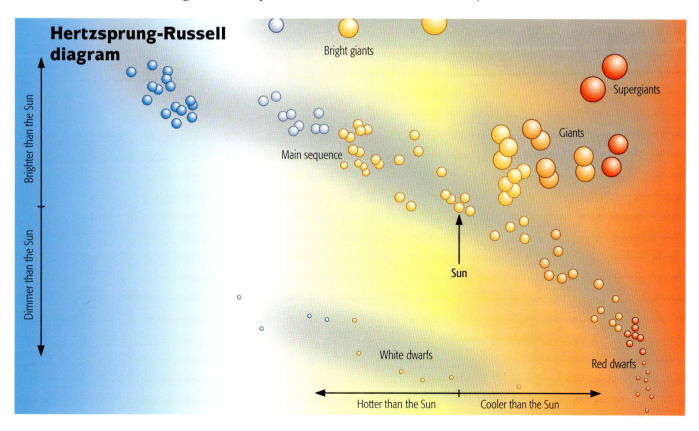

FOR MORE ON ORGANIZING THE HEAVENS SEE *REFRACTING TELESCOPES* 5:22; *REFLECTING TELESCOPES* 5:28; *RADIO TELESCOPES* 6:36

Organizing the heavens

sequence, which stretches diagonally from the top left to bottom right. But two groups of stars do not fall on the main sequence. One of these rogue groups, lying above the main sequence, consists of the red giant and supergiant stars, much brighter than might be expected from their color or temperature. Cooler stars, below the main sequence, are called white dwarfs.

The life story of a typical star can be traced on the Hertzsprung–Russell (or HR) diagram. A young star begins as a protostar, a cloud of gas and dust that condenses under its own gravity. When it is hot enough for fusion reactions to begin, it joins the main sequence. When all of its hydrogen has been consumed, the star first collapses, gets hotter, and then expands rapidly to become a red giant or even supergiant star. Finally, the star loses its outer layers (as a planetary nebula) and collapses into a white dwarf. The whole trip usually takes thousands of millions of years.

A red giant star may be the size of the Earth's orbit around the Sun; so if our Sun evolves into a giant, it will gobble up Mercury, Venus, and Earth as it expands. A white dwarf is tiny in star terms, being only about the size of the Earth. In a high-mass star the core may collapse explosively, and the star will throw off its outer layers as a brilliantly bright supernova. A simultaneous implosion inward crushes what remains of the core until it becomes a neutron star or a black hole, which represent the end of a star's life.

Gyrating galaxies

Stretching right across the middle of the night sky is a wide band of stars called the Milky Way. It is a galaxy—the galaxy that contains the Sun and many nearby stars. Beyond our galaxy there are millions of other galaxies, which look like smudgy patches of light in small telescopes. Only the most powerful telescopes can resolve them into millions of individual stars.

Elliptical galaxies

▲ **Edwin Hubble** classified galaxies by their shapes, which vary from spherical and elliptical to spiral or barred-spiral. It is thought that all galaxies are huge disks of stars, and their shape, "spherical" or "elliptical," depends on whether we see them face on (from above or below) or edge on. The disks slowly rotate and may develop spiral arms. Some spiral galaxies, the barred spirals, have a "bridge" of stars that stretches across the middle of the galaxy.

Galaxies can be classified according to their shapes, as was first done by the American astronomer Edwin Hubble in the 1930s. In his scheme, illustrated across these pages, there are three groups: elliptical galaxies, spiral galaxies, and barred-spiral galaxies. Despite these apparent differences, astronomers now think that all galaxies are essentially disk-shaped. Ellipticals consist of a disk of stars with a bulge at the center. If viewed edge on from Earth, the disk will have an elongated elliptical shape. The same galaxy viewed from above (or below) would look round, like a sphere. Spiral galaxies also have different appearances according to whether or not they are seen edge on. All are slowly rotating around their central nucleus.

Spiral galaxies

Barred-spiral galaxies

Organizing the heavens

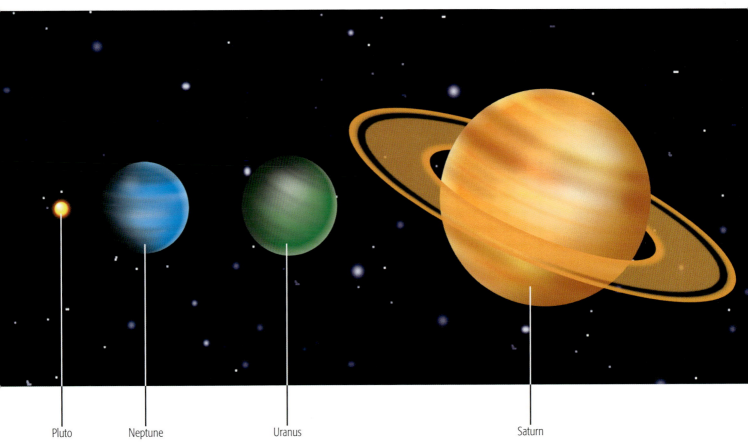

Pluto   Neptune   Uranus   Saturn

▲ **Most of the planets** fall into one of two major groups: the Earth-like terrestrial planets (Mercury, Venus, Earth, and Mars) and the gas giant planets (Jupiter, Saturn, Uranus, and Neptune). Tiny Pluto does not belong to either group. Here they are shown drawn to the same scale.

In addition to visible stars the heavens contain many other starlike objects. Radio stars give off high-frequency radio waves instead of light. Others pulsate, getting regularly brighter and dimmer. Still others emit x-rays. Black holes, the most mysterious, do not emit any radiation at all, yet astronomers know they are there from their effects on other nearby objects.

The Solar System

The Solar System consists of the Sun, nine planets (including Earth) in orbit around it, and lots of moons (including our Moon) that orbit the planets.

So far 61 planetary moons have been "measured" and given names. Also orbiting the Sun are thousands of asteroids or minor planets and comets that zoom in from outer space, swing around the Sun, and then zoom back out again. Associated with some of the comets are showers of meteors, or falling stars, seen each year at the same times.

Looking at their sizes (shown above in their correct relative sizes), there are two obvious groups of planets. There are the inner four—orbiting nearest the Sun—that include the Earth. They are all small rocky worlds with a density several times that of water. The four giant outer planets are not much denser than water (Saturn's density is even less). Most of their volume consists of gases. Distant Pluto is the smallest of all and so different that many astronomers question whether it should be included at all among the major planets.

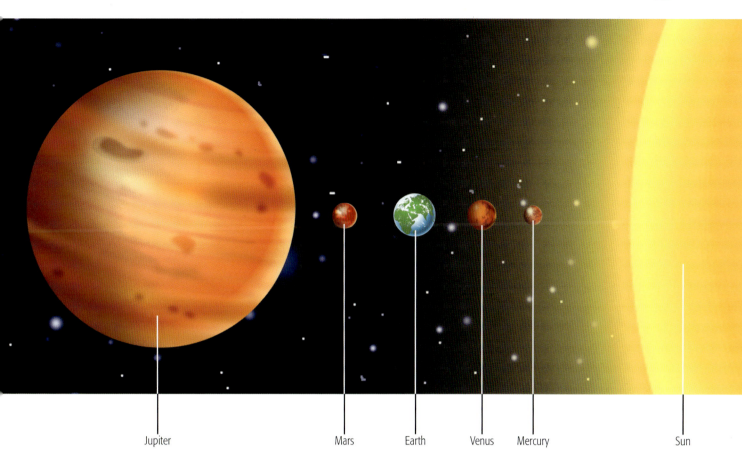

Jupiter   Mars   Earth   Venus Mercury   Sun

Edwin Hubble

Edwin Powell Hubble was born in 1889 in Marshfield, Missouri, the son of a lawyer. He trained in astronomy at Chicago University before going to Oxford, England, to study law. He worked briefly at Chicago's Yerkes Observatory from 1914 before joining the army in World War I (he was wounded while fighting in France), after which he joined the Mount Wilson Observatory in California. His main objects of study were the huge groups of stars called galaxies. In 1929 he discovered Hubble's law, which shows that the universe is expanding. He also devised a classification system for galaxies. He died in 1953. The Hubble Space Telescope, placed in orbit by the Space Shuttle in 1990, was named in his honor.

Glossary

Any of the words in SMALL CAPITAL LETTERS can be looked up in this Glossary.

angiosperm A type of plant that produces seeds that are enclosed within a FRUIT. See also GYMNOSPERM.

animal A many-celled living organism that gets its energy from eating plants or other animals. Most animals have nerves and senses, and can move around. Unlike PLANTS, their cell walls lack cellulose.

antibody A protein substance in the blood that combines with a particular ANTIGEN, such as a TOXIN produced by BACTERIA. This is how antibodies fight infection.

antigen A "foreign" substance, such as a TOXIN from BACTERIA, that stimulates the formation of ANTIBODIES when it enters the blood.

asteroid Also called a *minor planet*, any one of thousands of small, rocky bodies that orbit the Sun, mainly between the orbits of Mars and Jupiter.

atomic number An older name for PROTON NUMBER.

atomic weight An older name for RELATIVE ATOMIC MASS.

bacteria Microscopic organisms that can cause disease. They make up one of the two phyla in the kingdom of Monerans. (One such organism is a *bacterium*.)

binary star A type of STAR that consists of two stars orbiting each other.

binomial nomenclature The scientific way of naming organisms such as plants and animals. Each organism has a two-part name: the GENUS name followed by the SPECIES name.

black hole The object that results when a massive STAR at the end of its lifetime collapses completely in on itself because of its own gravity. Not even light can escape from the gravitational field of a black hole.

blood group Also called *blood type*, one of the four principal kinds (A, B, AB, and O) of human blood. They differ in whether there are particular ANTIGENS on the surface of the red blood cells.

class A classification grouping (e.g., for plants and animals) that consists of a set of related ORDERS. Related classes make up a DIVISION of plants or a PHYLUM of animals.

comet A small icy body in an orbit (usually highly elliptical) around the Sun.

constellation A group of stars that have been given a name used to identify a particular area of the night sky. Most of the stars in any one constellation are not physically close to one another—they just appear to be close when viewed from Earth.

cotyledon A seed leaf, the first part of a plant embryo (germinating seed) to emerge from the soil.

crystal habit The external shape of a crystal (e.g., cubic or hexagonal).

crystal structure The regular arrangement of atoms or atomic groupings within a crystal that gives it its shape.

Dewey decimal system A system of classifying books by subject used in many libraries.

dicotyledon A plant that has two COTYLEDONS (seed leaves).

division One of the 10 major groupings that together make up the plant kingdom. It is equivalent to a PHYLUM in animal classification.

dwarf star A "normal" star like the Sun on the main sequence of the HERTZSPRUNG–RUSSELL DIAGRAM, so named to distinguish them from RED GIANTS, which are many time larger and more luminous. See also WHITE DWARF.

family A classification grouping (e.g., for plants and animals) that consists of a set of related genera (plural of GENUS). Related families make up an ORDER.

fruit The part of a plant that surrounds the seeds.

galaxy A vast collection of stars, dust, and gas that may be *elliptical, spiral,* or *barred spiral* in shape.

genus A classification grouping (e.g., for plants and animals) that consists of a set of related SPECIES. Related *genera* (the plural of GENUS) make up a FAMILY.

Gram-negative A category of BACTERIA that do not (or only very weakly) absorb a dye called Gram's stain.

Gram-positive A category of BACTERIA that easily absorb a dye called Gram's stain.

group A vertical column of the PERIODIC TABLE that contains elements with similar physical and chemical properties.

gymnosperm A type of plant that bears naked seeds (i.e., not inside a fruit), usually in cones. See also ANGIOSPERM.

Hertzsprung–Russell diagram (HR diagram) A graph on which stars are positioned according to their brightness and temperature. Most stars lie on a diagonal line called the *main sequence.*

igneous rock A type of rock that originated in the Earth's molten magma before coming to the surface through fissures or volcanoes.

main sequence See HERTZSPRUNG–RUSSELL DIAGRAM.

metamorphic rock A type of rock originally formed under the surface from SEDIMENTARY ROCK (or sometimes IGNEOUS ROCK) by the action of heat and pressure. Subsequent weathering and earth movements may bring metamorphic rocks to the surface.

meteor Also called a *falling star,* a bright streak of light in the night sky formed when a tiny fragment of a COMET enters the Earth's atmosphere and burns up.

meteorite A rocky fragment of an asteroid that penetrates the Earth's atmosphere and strikes the ground.

Milky Way The GALAXY to which the Sun belongs; it is sometimes called the Galaxy, with a capital G.

mineral A naturally occurring chemical compound that always has the same composition wherever it is found. Several minerals may come together to form ROCK.

monocotyledon A plant that has only one COTYLEDON (seed leaf).

moon A natural satellite orbiting a planet. The Moon (with a capital M) is the Earth's natural satellite.

neutron star A type of dense, extremely small celestial object that is left behind after a massive STAR explodes as a SUPERNOVA.

order A classification grouping (e.g., for plants and animals) that consists of a set of related FAMILIES. Related orders make up a CLASS.

period A horizontal row of the PERIODIC TABLE that contains elements whose PROTON NUMBERS (atomic numbers) increase by 1 as you move along the row from left to right.

Periodic Table An arrangement of the chemical elements, in GROUPS and PERIODS according to their proton number, that reveals various similarities in their chemical properties.

photosynthesis The process by which green plants use the Sun's energy to convert water (from the soil) and carbon dioxide (from the air) into chemicals such as sugar and starch.

phylum One of the 14 major groupings that together make up the animal kingdom. It is equivalent to a DIVISION in plant classification.

planet Any one of the nine large celestial objects that orbit the Sun. See also ASTEROID; COMET.

plant A many-celled living organism that gets its energy from PHOTOSYNTHESIS. A plant has no nerves or senses and cannot move around. Its cell walls are made of cellulose.

proton number Formerly called *atomic number,* the number of protons in the nucleus of an atom, which gives an element its chemical identity.

red giant A type of large, cool star, typically tens of times as large as the Sun.

relative atomic mass Formerly called *atomic weight,* the average mass of an element's atom divided by one-twelfth of the mass of an atom of carbon-12.

rhesus factor An ANTIGEN on the surface of red blood cells that is present in people who are rhesus-positive (Rh+) but absent from people who are rhesus-negative (Rh–).

rock A naturally occurring hard substance that consists of a collection of MINERALS. Its composition may vary slightly depending on where it is found.

rock cycle The continuous process in which the products of WEATHERING of IGNEOUS ROCKS and METAMORPHIC ROCKS become SEDIMENTARY ROCKS, which in turn may form more

metamorphic rocks or become igneous rocks deep underground.

sedimentary rock A type of rock formed by the solidification of sediments of particles formed by the WEATHERING of IGNEOUS ROCKS (or sometimes METAMORPHIC ROCKS).

species A classification grouping (e.g., for plants and animals) that consists of organisms that can interbreed with one another. Related species make up a GENUS.

star A huge ball of gas that radiates light and other energy produced by nuclear fission reactions taking place within it. The main reaction converts hydrogen into helium.

supernova A spectacularly bright celestial object resulting from the explosion of a STAR, usually because it has run out of nuclear fuel.

taxonomic name An organism's biological name given by its GENUS and SPECIES. See BINOMIAL NOMENCLATURE.

taxonomy The science of biological classification.

toxin A poisonous substance produced by BACTERIA. If it enters the blood, a toxin stimulates the formation of ANTIBODIES.

universal donor A person who has type O blood, which will mix with any other blood group without harmful effects. See also BLOOD GROUP.

universal recipient A person who has type AB blood, who can be transfused with blood of any other group without any harmful effects. See also BLOOD GROUP.

weathering The action of wind, rain, ice, chemical change, and temperature change on rocks, which eventually breaks them down into small particles (such as sand).

white dwarf A type of small, dense STAR of low temperature and brightness. "Normal" stars end their lives as white dwarfs.

Further reading/websites and picture credits

Further Reading

Atoms and Molecules by Philip Roxbee-Cox; E D C Publications, 1992.

Electricity and Magnetism (Smart Science) by Robert Sneddon; Heinemann, 1999.

Electronic Communication (Hello Out There) by Chris Oxlade; Franklin Watts, 1998.

Energy (Science Concepts) by Alvin Silverstein et al.; Twenty First Century, 1998.

A Handbook to the Universe: Explanations of Matter, Energy, Space, and Time for Beginning Scientific Thinkers by Richard Paul; Chicago Review Press, 1993.

Heat (How Things Work Series) by Andrew Dunn; Thomson Learning, 1992.

How Things Work: The Physics of Everyday Life by Louis A. Bloomfield; John Wiley & Sons, 2001.

Introduction to Light: The Physics of Light, Vision and Color by Gary Waldman; Dover Publications, 2002.

Light and Optics (Science) by Allan B. Cobb; Rosen Publishing Group, 2000.

Electricity and Magnetism (Fascinating Science Projects) by Bobbi Searle; Copper Beech Books, 2002.

Basic Physics: A Self-Teaching Guide by Karl F. Kuhm; John Wiley & Sons, 1996.

Eyewitness Visual Dictionaries: Physics by Jack Challoner; DK Publishing, 1995.

Makers of Science by Michael Allaby and Derek Gjertsen; Oxford University Press, 2002.

Physics Matters by John O.E. Clark et al.; Grolier Educational, 2001.

Science and Technology by Lisa Watts; E D C/Usborne, 1995.

Sound (Make It Work! Science) by Wendy Baker, John Barnes (Illustrator); Two-Can Publishing, 2000.

Websites

Astronomy questions and answers — http://www.allexperts.com/getExpert.asp?Category=1360

Blood classification — http://sln.fi.edu/biosci/blood/types.html

Chemical elements — http://www.chemicalelements.com

Using and handling data — http://www.mathsisfun.com/data.html

Diffraction grating — http://hyperphysics.phy-astr.gsu.edu/hbase/phyopt/grating.html

How things work — http://rabi.phys.virginia.edu/HTW/

Pressure — http://ldaps.ivv.nasa.gov/Physics/pressure.html

About rainbows — http://unidata.ucar.edu/staff/blynds/rnbow.html

Story of the Richter Scale — http://www.dkonline.com/science/private/earthquest/contents/hall2.html

The rock cycle — http://www.schoolchem.com/rk1.htm

Views of the solar system — http://www.solarviews.com/eng/homepage.htm

The physics of sound — http://www.glenbrook.k12.il.us/gbssci/phys/Class/sound/u11l2c.html

A definition of mass spectrometry — http://www.sciex.com/products/about mass.htm

Walk through time. The evolution of time measurement — http://physics.nist.gov/GenInt/Time/time.html

How does ultrasound work? — http://www.imaginiscorp.com/ultrasound/index.asp?mode=1

X-ray astronomy — http://www.xray.mpe.mpg.de/

Picture Credits

Abbreviation: SPL Science Photo Library